# FASCINATING CIVIL ENGINEERING INFORMATION

## The Idea Guide for Young Minds

Dylan Austin

# SOME FASCINATING INTERIORS WITHIN

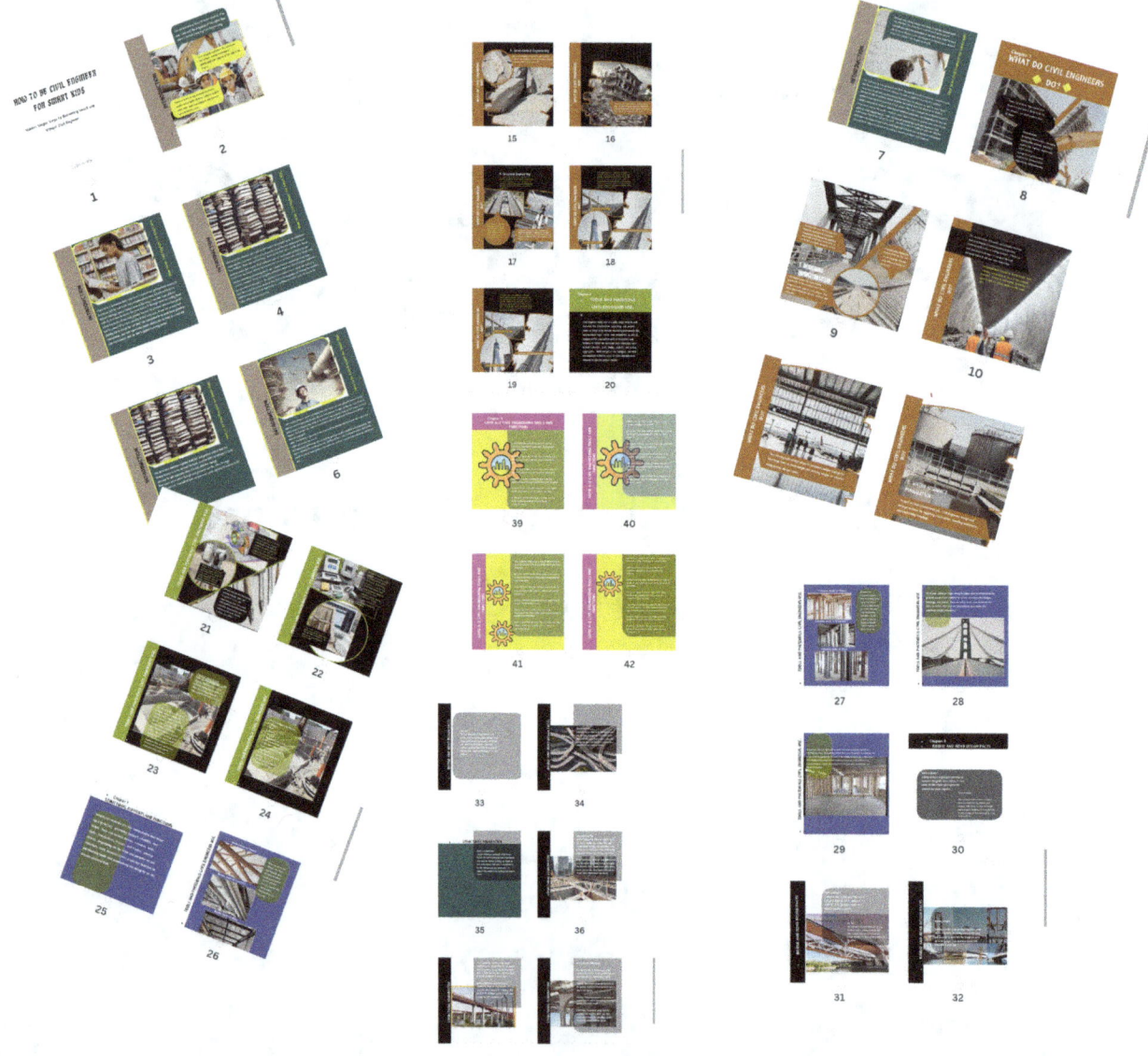

## INTRODUCTION

> Let me tell you a story of how I started. You see, I was not the brightest at the initial time when I started studying Civil Engineering

> I was so highly motivated. The enthusiasm was so high I wanted to become a professional Civil Engineer at the click of my fingers.

> I wanted to hit the top from the bottom so quickly and so badly. Only to realized I needed some work. Some real extensive and intensive work behind the scene.

1

## INTRODUCTION

Fascinating Civil Engineering Information

I have been a bookworm since i was teen. I remembered getting lunch money to buy books just to read in our noisy neighbourhood back in Eastern Market in Detroit, Michigan where I helped mom in the shop. I was able cope with the noise by learn how to silence the great deal of it through focus on reading those interesting books. It was only possible because the passion i had was too great I had to convert it to create serenity in that noisy area. It was not an easy feat starting out to accept the fact that i am surrounded by so much noise pollution. Professionals, and from experiential points advices one to study in a cool and calm environment. But this didn't applied to young Austin.

## INTRODUCTION

*Fascinating Civil Engineering Information*

Who knows the situation you might be in? I wasn't all brought up by an established single parent---my mom. I grew up to learn she's been the one taking care of us. Five Children. Three girls and two boys. And i am the first son--fourth child. Never grew up seeing dad around. That explained some of the struggles at that early age. I read a lot and books became my number one friend. At that time, not reading a day felt like something was missing out of me. I could feel the discomfort in my head. You see, inasmuch as we have negative addiction, there are tons of positive addiction one can become entangle in. The results of these addictions is usually the best of it kind. But i must warn you ahead of time that in the beginning of forming that positive addictions you will have to fight laziness, unwillingness not to read, the excuse to follow after pleasures and play.

## INTRODUCTION

Fascinating Civil Engineering Information

The difference between negative addiction and positive ones is that the former is easy to form, difficult to break. And the latter is difficult to form and easy to quit especially when it is not addictive enough.

One comes more often through easy pleasures and the other through intentional act. How intentional are you about your capacity to achieve in life young?

# INTRODUCTION

## Fascinating Civil Engineering Information

I still recalled when he came a particular time and found out i was going to study a course that was not in Engineering Faculty. He was not pleased. He said, young boy, i want you to pick a course in Engineering. Well, to be honest i was going to study an Art course not even Science. So how does he expect a young lad to pull out such stunt to become a Science student overnight?

When i heard about it [because it was convey to me through my mom that this is what dad said] I sat and ponder over it.

But what truly sparked my interest and imagination was the world of civil engineering. The idea of shaping the world around me, building bridges, and creating structures that would stand the test of time ignited a fire within me. And so, i accepted the challenge laid before me.

## INTRODUCTION

Oh boy! The science course was rocky. It was like moving uphill and downhill. The intensity of work to meet up the course demand was scary but fascinating at the same time. I had to learn the basics of Physics. I was already good at Mathematics. Because it was one subject I was taught constantly.

Fascinating Civil Engineering Information

My first victory over getting it was that i thought to myself in my head, "someone wrote the Physics textbook. It took a human to write it, right? It's not like it is spiritual Therefore, it will take another human to understand it. Because we got the same brain." Because at first be pattern Physics problem was solved were like matrix in my head. I couldn't wrap my head around it at first glance. But once I thought about that to myself, I found a new kind of sense of achievement bobbling within me--the momentum to carry on. If you have read to this point, I think it's about time I tell you to prepare your mind for the world of Civil Engineering as I walk you through fascinating facets and beauty of civil engineering. And at the end of the book, you will be able to familiarise yourself with the different endeavours carried out by Civil Engineers and how you can position and prepare yourself to be unique in pursuit of becoming a smart and vibrant Civil Engineer.

# Chapter 1
# WHAT DO CIVIL ENGINEERS ◆ DO? ◆

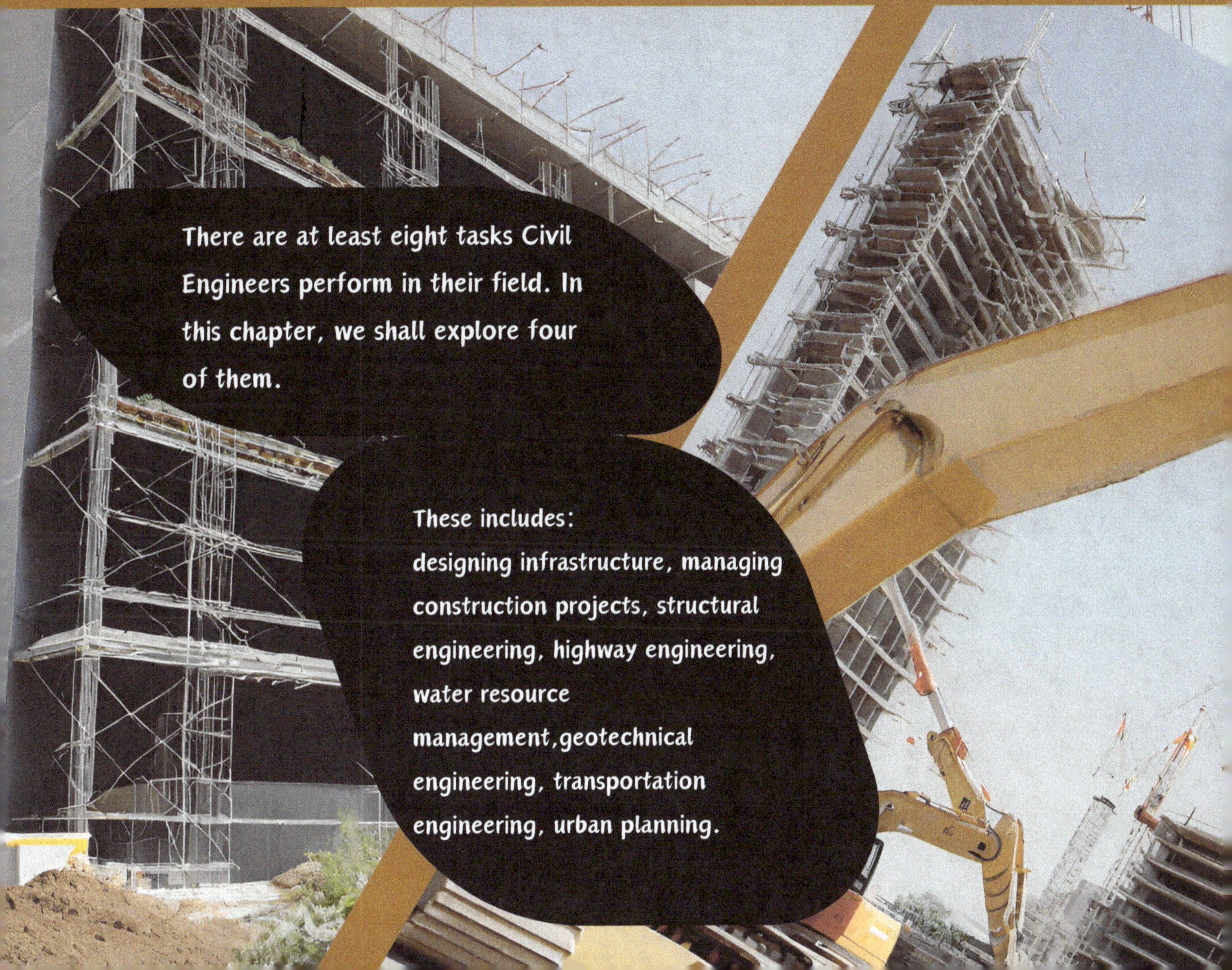

There are at least eight tasks Civil Engineers perform in their field. In this chapter, we shall explore four of them.

These includes: designing infrastructure, managing construction projects, structural engineering, highway engineering, water resource management, geotechnical engineering, transportation engineering, urban planning.

## 1. DESIGNING INFRASTRUCTURE

Civil Engineers are the professionals responsible for the construction of roads, bridges, tunnels, airports, and railways. They ensure these structures are safe, efficient, and sustainable.

They know the exact kind of bridge that should be constructed in a particular geographical area through data analytical means such as the kind of wheel loads or traffic that would pass through it overtime.

Transportation Engineers designs Railways that cut across impossible hurdles with different choices of materials.

The construction of Tunnels are so highly technical that Civil Engineers often involves innovative methods such as tunnel boring machines (TBMs) or drill and blast techniques to navigate through different types of soil and rock formations.

GOTTHARD BASE TUNNEL IN SWITZERLAND. THIS TUNNEL, COMPLETED IN 2016, IS THE LONGEST RAILWAY TUNNEL IN THE WORLD, SPANNING 57 KILOMETERS (35 MILES) BENEATH THE SWISS ALPS.

## WHAT DO CIVIL ENGINEERS DO?

## WHAT DO CIVIL ENGINEERS DO?

They strategically design Airports runaways unique enough to take huge tons of dead weight and moving weight of aircrafts using precise design codes

## WHAT DO CIVIL ENGINEERS DO?

## 2. WATER RESOURCE MANAGEMENT

As part of their professional job, Civil Engineers design and manage systems for supplying clean water, handling wastewater, and controlling flooding.

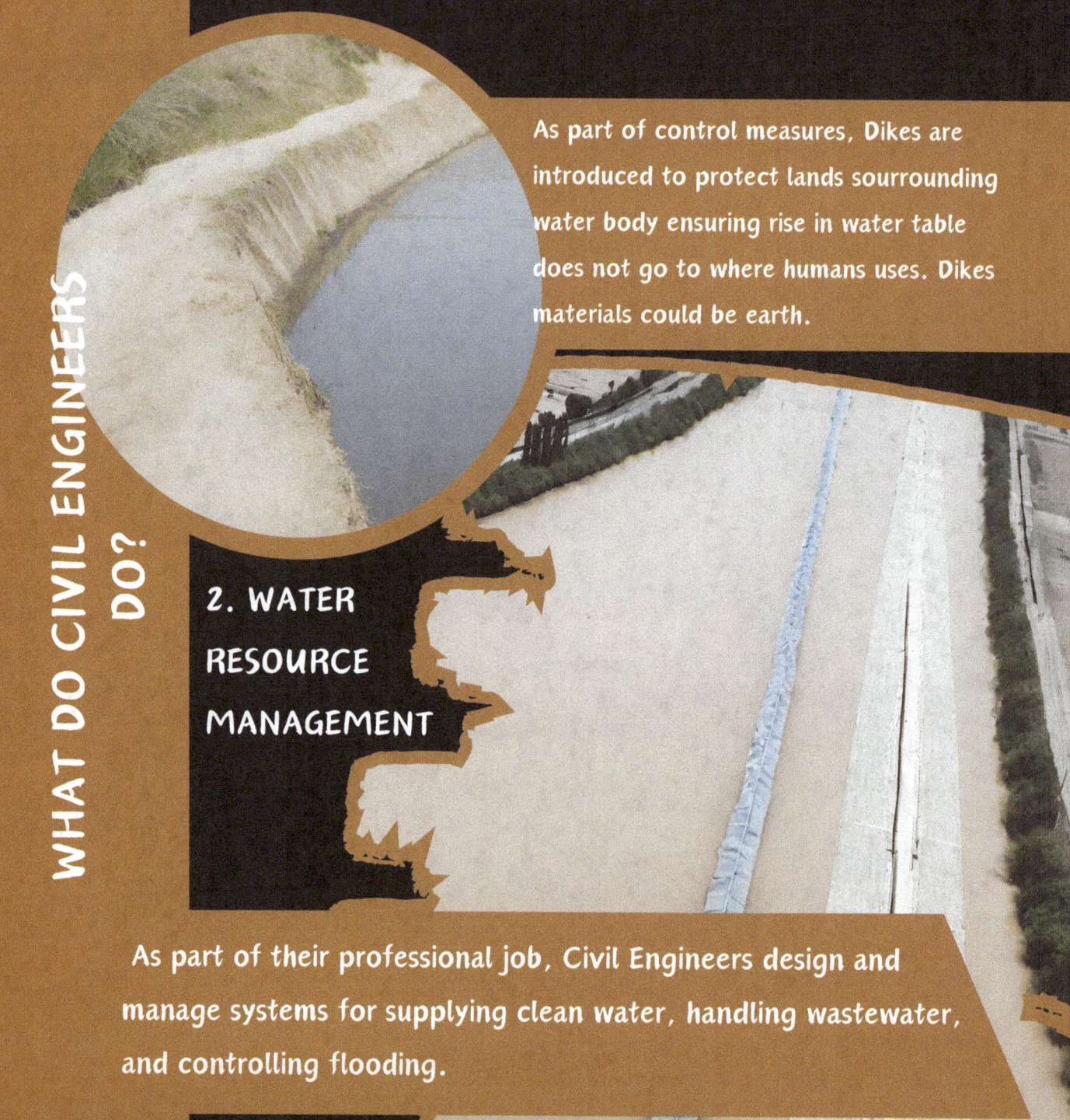

## WHAT DO CIVIL ENGINEERS DO?

### 2. WATER RESOURCE MANAGEMENT

As part of control measures, Dikes are introduced to protect lands sourrounding water body ensuring rise in water table does not go to where humans uses. Dikes materials could be earth.

As part of their professional job, Civil Engineers design and manage systems for supplying clean water, handling wastewater, and controlling flooding.

## WHAT DO CIVIL ENGINEERS DO?

Levees are quit similar with dikes. Levees are natural or man-made embankments or barriers designed to prevent water from overflowing its banks, usually along rivers, streams, or coastlines. They're built to protect communities and agricultural land from flooding.

Many are very familiar with Dams ability to produce electricity. But very few know that they are also part of flood control measures. The world recently have been facing series of flood. Some of these places lacks adequate Dams that can withhold the body of water released by human activities or nature

## WHAT DO CIVIL ENGINEERS DO?

## 3. Geotechnical Engineering

This involves studying the properties of soil and rock to determine their suitability for construction projects, such as building foundations and retaining walls.

## WHAT DO CIVIL ENGINEERS DO?

When it comes Geotech Engineering, Geotech engineers play a vital role in designing structures to withstand the disastrous earthquakes by understanding how different soil types and formations respond to seismic forces.

## WHAT DO CIVIL ENGINEERS DO?

## 4. Structural Engineering

Civil engineers design and analyze fascinating structures to ensure they can withstand the forces they will encounter during their lifespan, including gravity, wind, and earthquakes. The main thing is to design a static body that can remain immovable while holding a body of load including its own self weight coming upon it.

The Panama Canal, completed in 1914, remains one of the most impressive engineering feats in history. It allows ships to travel between the Atlantic and Pacific Oceans, saving thousands of miles of travel around South America.

The International Space Station (ISS) is a testament to international collaboration in engineering. It's the largest human-made structure in space, orbiting Earth at an average altitude of 420 kilometers and traveling at a speed of 28,000 kilometers per hour.

## WHAT DO CIVIL ENGINEERS DO?

Civil engineers typically use a combination of advanced structural engineering principles, innovative materials, and meticulous planning to build gravity-defying structures. This can involve techniques such as tension cables, cantilevers, counterweights, and strategic distribution of loads to achieve stability and balance despite gravity's force. Examples include skyscrapers, bridges, and unique architectural marvels like the Burj Khalifa or the Millau Viaduct.

Millau Viaduct structure in France.

World Trade Center-United State

The Burj Khalifa Tower standing tall in United Arab Emirates.

# Chapter 2
## TOOLS AND MATERIALS CIVIL ENGINEERS USE.

Civil engineer make use of a wide range of tools and materials for construction, surveying, and analysis. Some of these tools include measuring instruments like measurement tape, levels, and theodolites, as well as equipment for excavation such as excavators and bulldozers. While the materials they commonly used include concrete, steel, timber, asphalt, and various aggregates. These are just a few examples; the field encompasses a diverse array of tools and materials tailored to specific project needs.

## TOOLS AND MATERIALS CIVIL ENGINEERS USE.

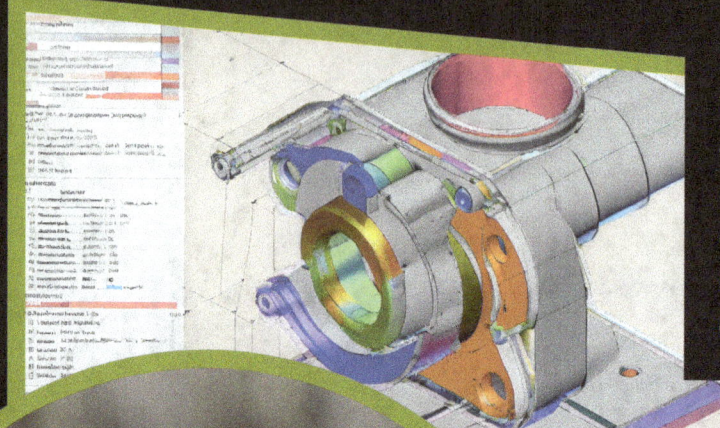

While drafting tools are used used for manual detailing, Computer aided designs such as AutoCAD, ArchiCAD are used to facilate this drawings more accurately, and rendering them to better explain in a layman way for the clients.

Surveying tools are crucial for accurately measuring and mapping land features. They include equipment like total stations, GPS receivers, levels, and theodolites. These tools help determine distances, angles, and elevations, essential for construction, engineering, and land development projects.

Dafting tools are used for creating detailed drawings, plans, and blueprints of various civil engineering projects. These tools help engineers accurately represent structures, infrastructure, and systems such as roads, bridges, buildings, and utilities.

## TOOLS AND MATERIALS CIVIL ENGINEERS USE.

Environmental monitoring equipment can include instruments like air quality monitors, water quality sensors, weather stations, soil analyzers, and remote sensing devices. These tools help track various environmental parameters such as pollution levels, temperature, humidity, radiation, and biodiversity, enabling researchers and authorities to assess and manage environmental conditions.

Structural analysis softwares are used to analysis calculate structural members such as beam, slab, column, column base etc... This is so the that the structure will be up to standard of usage.

## TOOLS AND MATERIALS CIVIL ENGINEERS USE.

Concrete is one of the most important materials civil engineers can't do without when it comes to structural engineering. Concrete is the mixture of Coarse aggregate, Fine aggregate, Cement, and water

Each one of these materials has a particular designed mix ratio to another. This is important because each element has its own ratio of porosity and as such cannot be given the same volume when mixing. That's why we have ratios like M20 1:2:4. Where 1 is for a head pan of Cement, 2 head of Fine Aggregate, and 4 head pan of Coarse Aggregate. Note that, the head pan measurement material might be different in other instances.

## TOOLS AND MATERIALS CIVIL ENGINEERS USE.

List of Civil Engineering Materials:
1. Concrete
2. Steel
3. Asphalt
4. Wood
5. Aggregates (sand, gravel)
6. Reinforcement bars (rebar)
7. Pipes (steel, PVC, concrete)
8. Geo-synthetics (geotextiles, geomembranes)
9. Masonry materials (bricks, blocks)
10. Waterproofing materials (membranes, sealants)

# Chapter 3
# STRUCTURAL ELEMENTS AND FUNCTIONS

Structural elements are the components that make up a structure, providing support, stability, and shape. They can include beams, columns, walls, slabs, foundations, trusses, and frames, among others, depending on the type and purpose of the structure. Each element serves a specific function in distributing loads and maintaining the integrity of the overall construction.

## TOOLS AND MATERIALS CIVIL ENGINEERS USE.

### Beams made of Timber

Beams are structural elements that takes load from the slab, distribute it with its members and send it down to columns. They are usual vertical columns whose logitudunals are on the x-axis.

### Beams made of Concrete

### Beams made of Steel

## TOOLS AND MATERIALS CIVIL ENGINEERS USE.

### Columns made of Timber

### Columns made of Concrete

### Columns made of Steel

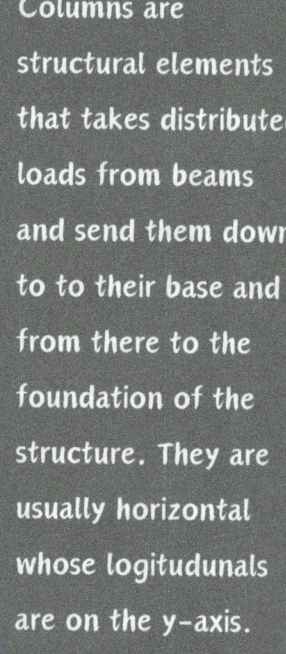

Columns are structural elements that takes distributed loads from beams and send them down to to their base and from there to the foundation of the structure. They are usually horizontal whose logitudunals are on the y-axis.

**TOOLS AND MATERIALS CIVIL ENGINEERS USE.**

Structural cables are high-strength cables used in construction to provide support and stability to various structures like bridges, buildings, and towers. They are often made from materials like steel or carbon fiber and are tensioned to bear loads and distribute weight effectively.

## TOOLS AND MATERIALS CIVIL ENGINEERS USE.

A structural slab is a reinforced concrete slab used to provide support to a building's structure. It's typically placed directly on the ground or on pilings and serves as the foundation for the rest of the building's construction. Structural slabs distribute the weight of the building evenly to prevent settlement or structural failure. They're commonly found in residential, commercial, and industrial buildings.

# BRIDGE AND ROAD DESIGN FACTS

**What's a Bridge?**

A bridge is like a magical path that helps us cross over things like rivers, valleys, or even roads! It's like a giant puzzle piece that connects two places together.

Types of Bridges

There are many different types of bridges. Some are shaped like big rainbows (arch bridges), while others are long and straight (beam bridges). Engineers use their creativity to design bridges that are strong and safe for everyone to use.

## What's a Bridge?

A bridge is like a magical path that helps us cross over things like rivers, valleys, or even roads! It's like a giant puzzle piece that connects two places together.

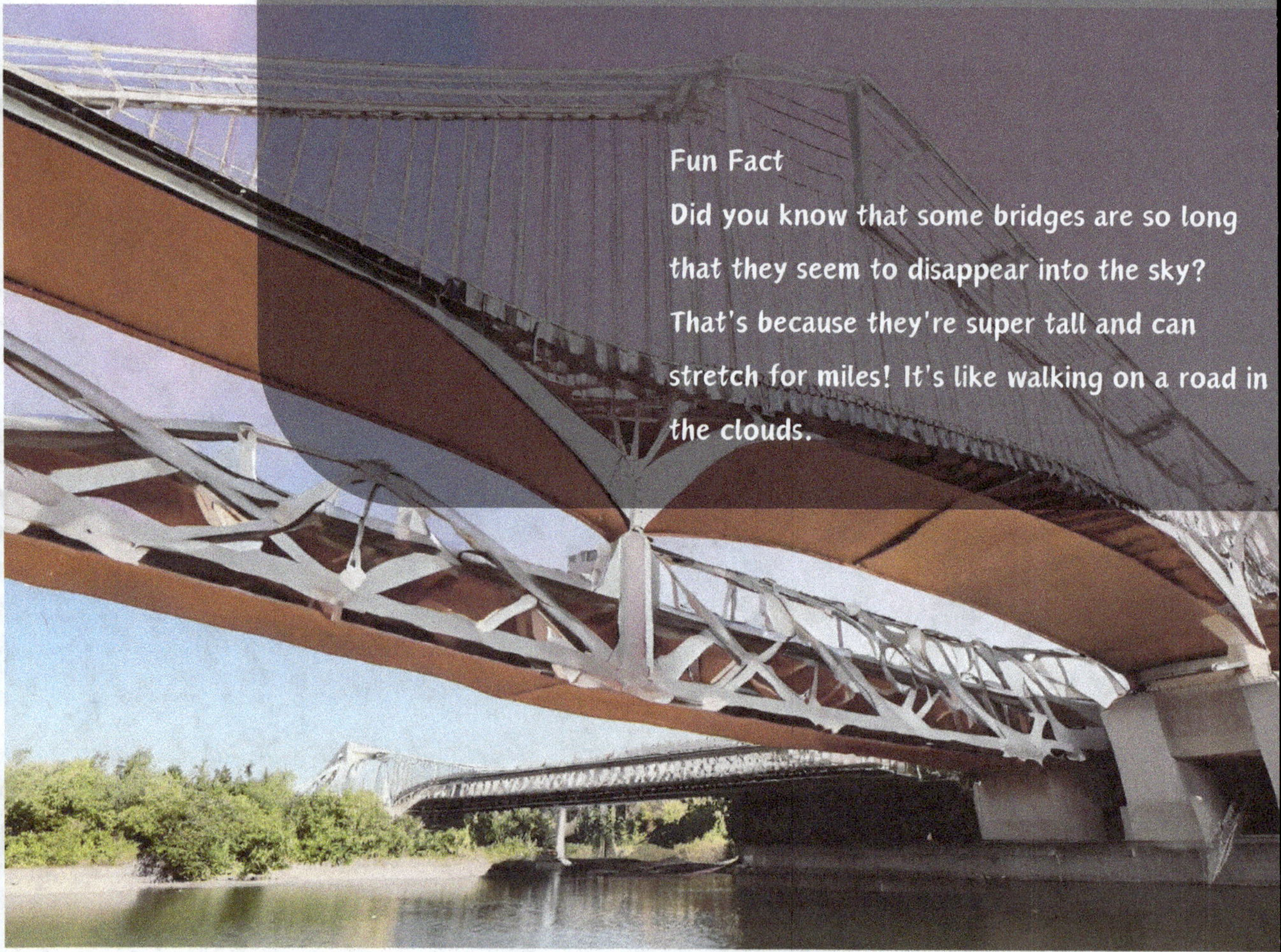

### Fun Fact
Did you know that some bridges are so long that they seem to disappear into the sky? That's because they're super tall and can stretch for miles! It's like walking on a road in the clouds.

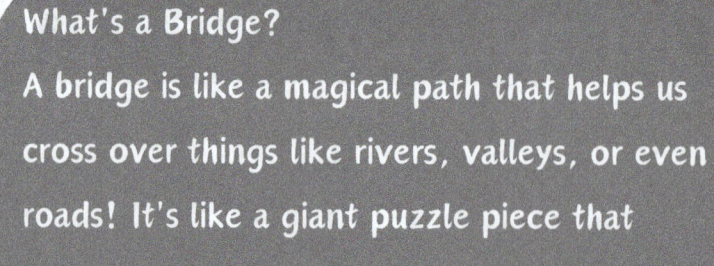

**BRIDGE AND ROAD DESIGN FACTS**

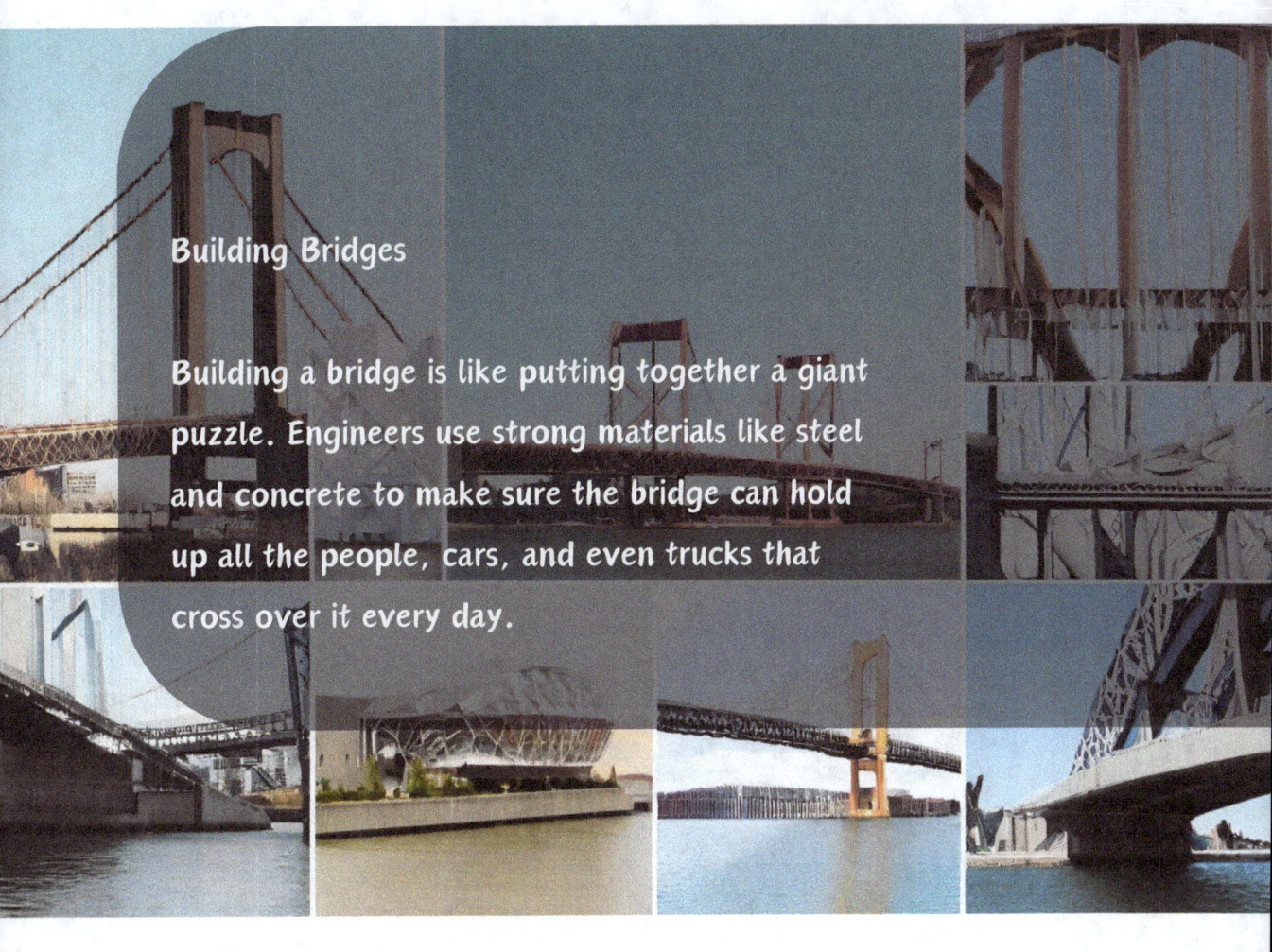

### Building Bridges

Building a bridge is like putting together a giant puzzle. Engineers use strong materials like steel and concrete to make sure the bridge can hold up all the people, cars, and even trucks that cross over it every day.

## BRIDGE AND ROAD DESIGN FACTS

Fun Fact

Did you know that the longest road in the world is over 48,000 kilometers (30,000 miles) long? That's like driving around the Earth and then some! It's called the Pan-American Highway, and it stretches from Alaska in North America all the way down to Argentina in South America.

# BRIDGE AND ROAD DESIGN FACTS

**Smooth and Safe**

Engineers design roads to be smooth and safe for everyone to travel on. They think about things like how many lanes to have, where to put traffic signs, and even how to make sure rainwater doesn't cause problems.

# STRUCTURAL FOUNDATION

What's a Foundation?

Imagine building a sandcastle at the beach. Before you start stacking up sand, you need to make sure the bottom is strong and stable so your castle doesn't fall down. A foundation is like the bottom layer of a sandcastle – it supports the weight of the building and keeps it steady.

## STRUCTURAL FOUNDATION

### FOUNDATION TYPES

**Shallow Foundations:** These are typically used for small to medium-sized structures. Types include spread footings, mat foundations, and slab-on-grade foundations. They distribute the load of the structure over a larger area of soil.

**Deep Foundations:** These are used when the soil near the surface isn't strong enough to support the structure's load. Types include piles, drilled shafts, and caissons. They transfer the load to deeper, more stable soil or rock layers.

## STRUCTURAL FOUNDATION

**Pile Foundations:** These are long, slender members driven or drilled deep into the ground. They can be made of concrete, steel, or timber. Piles are used when the soil conditions are poor at the surface but better deeper down.

**Raft Foundations:** Also known as mat foundations, these are large concrete slabs that cover the entire footprint of a building. They distribute the building's load over a wide area, suitable for soft or expansive soils.

## STRUCTURAL FOUNDATION

### FUNCTION OF FOUNDATION

**Load Distribution:** Foundations spread the weight of the structure evenly over the ground, preventing excessive settlement or failure.

**Stability:** Foundations anchor the structure to the ground, resisting overturning forces such as wind or earthquakes.

**Leveling:** Foundations provide a level base for constructing the rest of the building, ensuring it remains structurally sound.

**Protection:** Foundations can protect the structure from moisture, pests, and other environmental factors by providing a barrier between the building and the ground.

# Chapter 5
## SOME A-Z CIVIL ENGINEERING TOOLS AND FUNCTIONS

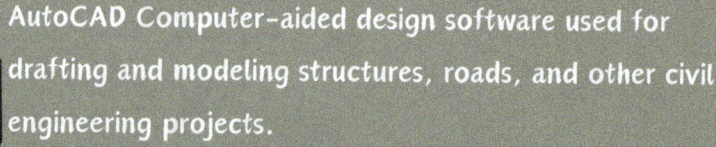

AutoCAD Computer-aided design software used for drafting and modeling structures, roads, and other civil engineering projects.

Bar Bender & Bar Cutter: Machines used to bend and cut reinforcement bars accurately according to construction needs.

Concrete Mixer: Equipment used to mix cement, water, and aggregate to produce concrete for construction projects.

Dumpy Level: An optical instrument used to establish horizontal lines and measure height differences in surveying.

Excavator: Heavy construction equipment used for digging trenches, foundations, and other earthmoving tasks.

GPS Receiver: Global Positioning System device used for precise location measurements in surveying and construction layout.

## SOME A–Z CIVIL ENGINEERING TOOLS AND FUNCTIONS

**Drill:** Power tool used for drilling holes in concrete, masonry, and other hard materials.

**Inclinometer:** Instrument used to measure angles of slope or tilt, helpful in monitoring stability of slopes and structures.

**Jackhammer:** Pneumatic tool used to break up concrete and rock during demolition or construction.

**Kneading Machine:** Equipment used for mixing soil and cement to create stabilized earth for construction.

**Laser Level:** Device that projects a straight horizontal or vertical line onto a surface, used for accurate leveling and alignment in construction.

**Manometer:** Instrument used to measure fluid pressure in hydraulic systems, vital for testing pipelines and water distribution systems.

## SOME A–Z CIVIL ENGINEERING TOOLS AND FUNCTIONS

**Noise Dosimeter:** Device used to measure and assess noise levels in construction sites to ensure compliance with safety regulations.

**Optical Theodolite:** Precision instrument used in surveying to measure horizontal and vertical angles for mapping and construction layout.

**Pneumatic Roller:** Heavy construction equipment used for compacting soil, asphalt, and other materials during road construction.

**Quicklime:** Chemical compound used in soil stabilization to improve its load-bearing capacity.

**Rebar Locator:** Tool used to detect and locate reinforcement bars (rebar) within concrete structures during inspections and renovations.

**Survey Total Station:** Electronic device used for measuring distances, angles, and coordinates in surveying and construction layout.

# SOME A-Z CIVIL ENGINEERING TOOLS AND FUNCTIONS

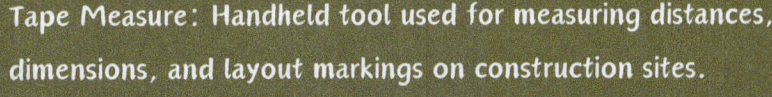

**Tape Measure:** Handheld tool used for measuring distances, dimensions, and layout markings on construction sites.

**Utility Knife:** Handy tool used for cutting various construction materials like plastic, cardboard, and insulation.

**Vibratory Plate Compactor:** Machine used for compacting soil, gravel, and asphalt surfaces to achieve greater density and stability.

**Water Level:** Simple instrument used for determining elevations and establishing level reference points in construction and landscaping.

**X-Ray Fluorescence (XRF) Analyzer:** Portable device used for on-site analysis of materials' chemical composition, helpful in quality control and contamination testing.

**Yardstick:** Long ruler used for measuring and marking straight lines on construction materials and layouts.

**Zinc Anode:** Sacrificial metal used to protect steel structures from corrosion in marine and underground environments.

# BASIC FORMULARS AND FUNCTIONS

Structural Analysis Formulas:

- Stress ($\sigma$): Force applied per unit area. Formula: $\sigma = F/A$
  - F: Force applied
  - A: Cross-sectional area
  - Example: If a force of 1000 N is applied to a steel rod with a cross-sectional area of 0.005 m², the stress is $\sigma$ = 1000 N / 0.005 m² = 200,000 N/m².
- Strain ($\varepsilon$): Deformation per unit length. Formula: $\varepsilon = \Delta L / L$
  - $\Delta L$: Change in length
  - L: Original length
  - Example: If a steel rod of length 2 m stretches to 2.1 m under a load, the strain is $\varepsilon$ = (2.1 m - 2 m) / 2 m = 0.05 or 5%.

Flexural Stress ($\sigma$): Stress due to bending moment. Formula: $\sigma = M*y/I$
  - M: Bending moment
  - y: Distance from neutral axis to the point of interest
  - I: Moment of inertia
  - Example: If a beam is subjected to a bending moment of 100 kN-m and the distance from the neutral axis to the extreme fiber is 0.2 m, with a moment of inertia of 0.01 m^4, the flexural stress is $\sigma$ = (100 kN-m * 0.2 m) / 0.01 m^4 = 2000 kN/m^2.
- **Deflection ($\delta$)**: Displacement of a point on a structural element under load. Formula: $\delta = (5WL^4) / (384EI)$
  - W: Applied load
  - L: Length of the beam
  - E: Modulus of elasticity
  - I: Moment of inertia
  - Example: If a simply supported beam with a span of 4 m carries a uniformly distributed load of 10 kN/m, with a modulus of elasticity of 200 GPa and a moment of inertia of 0.002 m^4, the deflection at the center is $\delta$ = (5 * 10 kN/m * (4 m)^4) / (384 * 200 GPa * 0.002 m^4) = 0.0052 m.

**BASIC FORMULARS AND FUNCTIONS**

Shear Stress (τ): Stress parallel to the cross-sectional area. Formula: $\tau = F/A$

- F: Force applied parallel to the surface
- A: Cross-sectional area parallel to the force
- Example: If a force of 500 N is applied parallel to the top surface of a rectangular block with dimensions 0.1 m x 0.2 m, the shear stress is $\tau = 500\ N / (0.1\ m * 0.2\ m) = 2500\ N/m^2$.

Torsional Stress (τ): Stress caused by twisting of a structural member. Formula: $\tau = (T * r) / J$

- T: Applied torque
- r: Distance from the center to the point of interest
- J: Polar moment of inertia
- Example: If a shaft is subjected to a torque of 100 N-m and has a radius of 0.05 m with a polar moment of inertia of 0.0005 m^4, the torsional stress is $\tau = (100\ N\text{-}m * 0.05\ m) / 0.0005\ m^4 = 10{,}000\ N/m^2$.

## BASIC FORMULARS AND FUNCTIONS

Darcy-Weisbach Equation: Calculates pressure loss due to friction in a pipe. Formula: $\Delta P = f(L/D)(\rho v^2)/2$

 - $\Delta P$: Pressure loss
 - f: Darcy friction factor
 - L: Length of pipe
 - D: Diameter of pipe
 - $\rho$: Density of fluid
 - v: Velocity of fluid
 - Example: If water is flowing through a pipe with a length of 100 m, diameter of 0.5 m, velocity of 2 m/s, and a friction factor of 0.02, the pressure loss would be $\Delta P = 0.02 * (100\ m / 0.5\ m) * (1000\ kg/m^3 * (2\ m/s)^2) / 2$.

Geotechnical Engineering Formulas:

Bearing Capacity: Maximum load a soil can support. Formula: $q = cN_c + \gamma D_f N_q + 0.5\gamma B N_\gamma$

 - q: Bearing capacity
 - c: Cohesion
 - $\gamma$: Unit weight of soil
 - Df: Depth factor
 - Nc, Nq, N$\gamma$: Bearing capacity factors
 - Example: For a cohesive soil with a cohesion of 30 kN/m², a unit weight of 18 kN/m³, and a depth factor of 1.2, calculate the bearing capacity.

## QUESTIONS AND ANSWERS

1. Question: What is civil engineering?

   Answer: Civil engineering is a branch of engineering that deals with the design, construction, and maintenance of infrastructure projects such as buildings, roads, bridges, dams, and water supply systems.

2. Question: What are the primary disciplines within civil engineering?

   - Answer: The primary disciplines within civil engineering include structural engineering, geotechnical engineering, transportation engineering, environmental engineering, and water resources engineering.

3. Question: What is the purpose of a structural load analysis?

   - Answer: Structural load analysis determines the magnitude and distribution of loads acting on a structure. It is essential for designing safe and economical structures.

4. Question: How do you calculate the dead load of a structure?

   - Answer: The dead load of a structure is calculated by summing the weights of all permanent components such as walls, floors, and roofs.

## QUESTIONS AND ANSWERS

5. Question: Define wind load and how it is calculated in structural design.

   - Answer: Wind load is the force exerted by the wind on a structure. It is calculated using wind speed, exposure category, and building dimensions according to relevant building codes or standards.

6. Question: What is the purpose of a factor of safety in structural design?

   - Answer: The factor of safety provides a margin of safety to account for uncertainties and variations in materials, construction, and loading conditions. It ensures structural reliability by comparing the actual load to the design load, with a safety margin.

7. Question: How do you calculate the moment of inertia of a beam?

   - Answer: The moment of inertia (I) of a beam depends on its shape. For example, for a rectangular beam, the formula is $I = \frac{bh^3}{12}$, where $b$ is the width and $h$ is the height of the beam.

8. **Questi: Explain the concept of bending stress in a beam.

   - Answer: Bending stress in a beam is the internal stress induced by bending moments, causing tension on one side and compression on the other. It is calculated using the formula $\sigma = \frac{My}{I}$, where $M$ is the bending moment, $y$ is the distance from the neutral axis, and $I$ is the moment of inertia.

9. Question: Define hydrostatic pressure and how it is calculated.

   - Answer: Hydrostatic pressure is the pressure exerted by a fluid at rest due to its weight. It is calculated using the formula $P = \rho \cdot g \cdot h$, where $\rho$ is the fluid density, $g$ is the acceleration due to gravity, and $h$ is the depth.

**QUESTIONS AND ANSWERS**

10. Question: What is the purpose of soil bearing capacity in foundation design?

    - Answer: Soil bearing capacity is the maximum pressure a soil can safely support without failure. It is essential in foundation design to ensure the stability and safety of structures.

11. Question: How do you calculate the bearing capacity of soil using Terzaghi's bearing capacity equation?

    - Answer: Terzaghi's bearing capacity equation is:
    $$q = cN_c + \gamma D_f N_q + \frac{1}{2} \gamma B N_\gamma$$

12. Question: Explain the concept of consolidation settlement in soil mechanics.

    - Answer: Consolidation settlement is the gradual compression and settlement of soil due to the expulsion of pore water under applied loads. It is caused by the rearrangement of soil particles and is time-dependent.

13. Question: What is the purpose of a drainage system in civil engineering projects?

    - Answer: A drainage system is designed to remove excess water from roads, buildings, and other structures to prevent waterlogging, erosion, and damage to the infrastructure.

14. Questio: How do you calculate the flow rate in a pipe using Manning's equation?

    - Answer: Manning's equation for flow rate in an open channel is:
    $$Q = \frac{1.486}{n} \cdot A \cdot R^{2/3} \cdot S^{1/2}$$

**QUESTIONS AND ANSWERS**

10. Question: What is the purpose of soil bearing capacity in foundation design?

    - Answer: Soil bearing capacity is the maximum pressure a soil can safely support without failure. It is essential in foundation design to ensure the stability and safety of structures.

11. Question: How do you calculate the bearing capacity of soil using Terzaghi's bearing capacity equation?

    - Answer: Terzaghi's bearing capacity equation is:
    $$ q = cN_c + \gamma D_f N_q + \frac{1}{2} \gamma B N_\gamma $$

12. Question: Explain the concept of consolidation settlement in soil mechanics.

    - Answer: Consolidation settlement is the gradual compression and settlement of soil due to the expulsion of pore water under applied loads. It is caused by the rearrangement of soil particles and is time-dependent.

13. Question: What is the purpose of a drainage system in civil engineering projects?

    - Answer: A drainage system is designed to remove excess water from roads, buildings, and other structures to prevent waterlogging, erosion, and damage to the infrastructure.

14. Question: How do you calculate the flow rate in a pipe using Manning's equation?

    - Answer: Manning's equation for flow rate in an open channel is:
    $$ Q = \frac{1.486}{n} \cdot A \cdot R^{2/3} \cdot S^{1/2} $$

15. Question: Define critical flow in open channel hydraulics.

    - Answer: Critical flow is the flow condition in an open channel where the Froude number equals unity, indicating a transition between subcritical and supercritical flow regimes.

16. Question: What is the formula for calculating the volume of water in a cylindrical tank?

    - Answer: The formula to calculate the volume of water in a cylindrical tank is:
    $$ V = \pi r^2 h $$

17. Question: How do you calculate the force exerted by water pressure on a submerged surface?

    - Answer: The force exerted by water pressure on a submerged surface is calculated using the formula:
    $$ F = P \cdot A $$

18. Question: What is the formula for calculating the flow velocity in a pipe based on the volumetric flow rate?

    - Answer: The flow velocity in a pipe can be calculated using the formula:
    $$ v = \frac{Q}{A} $$

QUESTIONS AND ANSWERS

## QUESTIONS AND ANSWERS

**20. Question:** What is the formula for calculating the pressure drop in a pipe due to friction?

   - **Answer:** The pressure drop in a pipe due to friction can be calculated using the Darcy-Weisbach equation:
   $$ \Delta P = f \cdot \frac{L}{D} \cdot \frac{\rho \cdot v^2}{2} $$

**21. Question:** Define shear stress in civil engineering and provide its formula.

   - **Answer:** Shear stress is the stress parallel to the cross-sectional area. It is calculated using the formula: $ \tau = \frac{F}{A} $, where $ F $ is the force applied parallel to the surface, and $ A $ is the cross-sectional area parallel to the force.

**22. Question:** What is the purpose of slope stability analysis in geotechnical engineering?

   - **Answer:** Slope stability analysis assesses the stability of slopes to prevent landslides and slope failures. It considers factors such as soil properties, groundwater conditions, and slope geometry.

23. Question: How do you calculate the hydraulic gradient in a pipe?

   - Answer: The hydraulic gradient is calculated using the formula: $$ i = \frac{\Delta h}{L} $$, where $i$ is the hydraulic gradient, $\Delta h$ is the change in hydraulic head, and $L$ is the length of the pipe.

24. Question: Define lateral earth pressure and provide its formula.

   - Answer: Lateral earth pressure is the pressure exerted by soil against retaining structures such as retaining walls. It is calculated using formulas like Rankine's theory or Coulomb's theory.

25. Question: What is the formula for calculating the volume of a rectangular prism?

   - Answer: The formula to calculate the volume of a rectangular prism is: $$ V = l \times w \times h $$, where $l$ is the length, $w$ is the width, and $h$ is the height of the prism.

26. Question: Define modulus of elasticity (Young's modulus) and provide its formula.

   - Answer: Modulus of elasticity (Young's modulus) is a measure of a material's stiffness. It is calculated using the formula: $$ E = \frac{\sigma}{\epsilon} $$, where $ E $ is the modulus of elasticity, $ \sigma $ is stress, and $ \epsilon $ is strain.

27. Question: What is the purpose of soil compaction in construction projects?

   - Answer: Soil compaction increases the density of soil to improve its load-bearing capacity, reduce settlement, and enhance stability. It is commonly used in road construction, foundations, and embankments.

28. Question: How do you calculate the area of a circle?

   - Answer: The area of a circle is calculated using the formula: $$ A = \pi r^2 $$, where $ r $ is the radius of the circle.

29. Question: Define coefficient of permeability in geotechnical engineering.

   - Answer: Coefficient of permeability (also known as hydraulic conductivity) is a measure of a soil's ability to transmit fluids. It represents the rate at which water can flow through the soil under a hydraulic gradient.

30. Question: What is the formula for calculating the settlement of a shallow foundation?

   - Answer: The settlement of a shallow foundation can be estimated using empirical methods such as the Terzaghi's formula: $$ \Delta S = \frac{\Delta \sigma}{1 + e_0} $$, where $\Delta S$ is the settlement, $\Delta \sigma$ is the increase in stress, and $e_0$ is the initial void ratio of the soil.

31. Question: Define truss in structural engineering.

   - Answer: A truss is a structure composed of straight members connected at joints. It is designed to support loads primarily by axial forces in its members.

32. **Question**: What is the formula for calculating the centroid of a geometric shape?

   - Answer: The centroid of a geometric shape can be calculated using integration techniques for complex shapes or geometric formulas for simple shapes like triangles, rectangles, and circles.

33. Question: Define porosity in soil mechanics.

   - Answer: Porosity is a measure of the void spaces (pores) in a soil sample. It indicates the volume percentage of voids to the total volume of the soil.

34. Question: How do you calculate the flow velocity in an open channel using the Manning's equation?

   - Answer: Manning's equation for flow velocity in an open channel is: $$ v = \frac{1}{n} R^{2/3} S^{1/2} $$, where $v$ is the flow velocity, $n$ is Manning's roughness coefficient, $R$ is the hydraulic radius, and $S$ is the slope of the channel.

35. Question: Define moment of resistance in structural engineering.

   - Answer: Moment of resistance is the maximum moment a structural member can withstand before failure. It is determined by the material properties and cross-sectional geometry of the member.

36. Question: What is the purpose of a geotechnical investigation?

   - Answer: A geotechnical investigation assesses subsurface soil and rock conditions to provide data for foundation design, slope stability analysis, and construction planning.

37. Question: How do you calculate the volume of water stored behind a dam?

   - Answer: The volume of water stored behind a dam is calculated by multiplying the cross-sectional area of the reservoir by the length of the reservoir.

38. Question: Define buoyancy in fluid mechanics.

   - Answer: Buoyancy is the upward force exerted on an object immersed in a fluid (liquid or gas) due to the difference in pressure between the top and bottom of the object. It is governed by Archimedes' principle.

39. Question: What is the formula for calculating the lateral earth pressure on a retaining wall?

   - Answer: The lateral earth pressure on a retaining wall can be calculated using theories such as Rankine's theory or Coulomb's theory, depending on the soil properties and wall geometry.

40. Question: Define the concept of traffic volume in transportation engineering.

   - Answer: Traffic volume refers to the number of vehicles passing through a given section of road or highway during a specified period, usually measured in vehicles per day (VPD) or vehicles per hour (VPH).

41. Question: How do you calculate the angle of repose of a soil?

   - Answer: The angle of repose of a soil is the maximum angle at which it remains stable without collapsing. It can be determined experimentally by slowly pouring the soil onto a flat surface until it forms a pile, and then measuring the angle of the pile from the horizontal.

42. Question: Define creep in structural engineering.

   - Answer: Creep is the gradual deformation of a material under a constant load over time. It is particularly important in materials such as concrete and soil, where long-term loading can cause significant deformation.

43. Question: What is the formula for calculating the moment of inertia of a circular section?

   - Answer: The moment of inertia (I) of a circular section can be calculated using the formula: $$ I = \frac{\pi}{64} \times (D^4 - d^4) $$, where $D$ is the outer diameter and $d$ is the inner diameter of the circular section.

44. Question: Define bearing capacity in geotechnical engineering.

   - Answer: Bearing capacity is the maximum pressure that a soil can support without failure. It is crucial for the design of foundations and other structures resting on the soil.

**QUESTIONS AND ANSWERS**

45. Question: How do you calculate the deflection of a beam under a given load?

   - Answer: The deflection of a beam under a given load can be calculated using structural analysis methods such as the moment-area method, virtual work method, or finite element analysis (FEA).

46. Question: Define poisson's ratio in material science.

   - Answer: Poisson's ratio is the ratio of lateral strain to longitudinal strain in a material subjected to an axial load. It is a measure of the material's deformation behavior under stress.

47. Question: What is the purpose of a geosynthetic material in civil engineering?

   - Answer: Geosynthetic materials, such as geotextiles, geogrids, and geomembranes, are used in civil engineering for various applications including soil stabilization, erosion control, drainage, and environmental protection.

48. Question: How do you calculate the volume of a trapezoidal footing?

   - Answer: The volume of a trapezoidal footing can be calculated by multiplying the average area of the trapezoid by the length of the footing.

49. Question: Define permeability in fluid mechanics.

   - Answer: Permeability is a measure of a material's ability to allow fluids to pass through it. It is an important property in groundwater flow and filtration processes.

50. Question: What is the formula for calculating the flow rate in a pipe based on the pressure difference?

   - Answer: The flow rate in a pipe based on the pressure difference can be calculated using Bernoulli's equation and the Darcy-Weisbach equation for head loss. The formula is often solved using numerical methods or flow meters in practical applications.

51. Question: Define soil compaction and its significance in construction.

   - Answer: Soil compaction is the process of increasing the density of soil by removing air voids. It is crucial in construction to improve soil strength, stability, and reduce settlement.

52. Question: How do you calculate the vertical stress at a point below the surface due to a uniformly distributed load?

   - Answer: The vertical stress ($\sigma_v$) at a point below the surface due to a uniformly distributed load can be calculated using the Boussinesq equation:
   $$\sigma_v = \frac{P}{2\pi z^2} \left(1 + \frac{d}{z}\right)$$

53. Question: Define composite construction in structural engineering.

   - Answer: Composite construction involves combining two or more materials (such as steel and concrete) to form a single structural element. It utilizes the strengths of each material to create efficient and economical structures.

54. Question: What is the formula for calculating the deflection of a simply supported beam under a point load?

   - Answer: The deflection (\( \delta \)) of a simply supported beam under a point load can be calculated using the formula for a concentrated load at the center:
   \[ \delta = \frac{PL^3}{48EI} \]

55. Question: Define coefficient of consolidation in soil mechanics.

   - Answer: Coefficient of consolidation (\( C_v \)) is a measure of the rate at which saturated clay soils consolidate under load. It is an important parameter for estimating settlement.

56. Question: How do you calculate the total head of a pump in a fluid system?

   - Answer: The total head (\( H \)) of a pump in a fluid system is the sum of the elevation head, pressure head, and velocity head:
   \[ H = z + \frac{P}{\rho g} + \frac{v^2}{2g} \]

**57. Question:** Define geogrid and its application in civil engineering.

- **Answer:** Geogrid is a geosynthetic material used for soil reinforcement and stabilization. It is commonly used in retaining walls, slopes, embankments, and road construction to improve soil strength and prevent erosion.

**58. Question:** What is the formula for calculating the modulus of subgrade reaction for a flexible pavement?

- **Answer:** The modulus of subgrade reaction ($k$) for a flexible pavement can be calculated using the Westergaard's equation:

$$ k = \frac{2E_h}{\left(1 - \nu_h\right)\left(1 + \nu_h\right)} $$

**59. Question:** Define seepage in soil mechanics and its significance in civil engineering.

- **Answer:** Seepage is the flow of water through soil or porous materials. It is significant in civil engineering for analyzing groundwater flow, seepage control, and stability of earth structures such as dams and levees.

60. Question: How do you calculate the area of a trapezoid?

- Answer: The area ($A$) of a trapezoid can be calculated using the formula:
$$ A = \frac{1}{2}(b_1 + b_2)h $$

61. Question: Define shear strength in soil mechanics.

- Answer: Shear strength is the maximum resistance of soil to shearing forces. It is a critical parameter for analyzing slope stability, bearing capacity, and foundation design.

62. Question: What is the formula for calculating the critical depth in open channel flow?

- Answer: The critical depth ($y_c$) in open channel flow can be calculated using the specific energy equation:
$$ y_c = \left(\frac{Q^2}{g}\right)^{1/3} $$

63. Question: Define traffic engineering and its objectives.

- Answer: Traffic engineering involves the planning, design, and operation of transportation systems to achieve safe, efficient, and sustainable traffic flow. Its objectives include reducing congestion, improving safety, and enhancing mobility.

64. Question: How do you calculate the vertical stress increase in soil due to a uniformly loaded area?

- Answer: The vertical stress increase ($\Delta \sigma_v$) in soil due to a uniformly loaded area can be calculated using the Boussinesq equation:

$$\Delta \sigma_v = \frac{q}{\pi} \left(\frac{1}{z_1} - \frac{1}{z_2}\right)$$

65. Question: What is the purpose of a geotechnical report?

- Answer: A geotechnical report provides detailed information about subsurface soil conditions, geology, groundwater levels, and recommendations for foundation design, slope stability, and construction methods.

66. Question: Define ductility in structural engineering.

   - Answer: Ductility is the ability of a material to deform plastically without fracture under applied stress. It is an important property for structures subjected to dynamic loads or seismic events.

67. Question: What is the formula for calculating the discharge coefficient for an orifice flow?

   - Answer: The discharge coefficient ($C_d$) for an orifice flow can be calculated using empirical correlations or obtained from experimental data.

68. Question: Define composite material and provide examples.

   - Answer: A composite material is a combination of two or more distinct materials with different properties to form a new material with enhanced performance. Examples include fiber-reinforced plastics, concrete with steel reinforcement, and laminated glass.

69. Question: How do you calculate the angle of internal friction for a granular soil?

   - Answer: The angle of internal friction ($\phi$) for a granular soil can be determined from triaxial or direct shear tests in the laboratory or estimated from empirical correlations based on soil properties.

70. Question: Define seismic design in structural engineering.

   - Answer: Seismic design involves designing structures to resist earthquake forces and minimize damage during seismic events. It considers factors such as building materials, structural configuration, and local seismic hazard.

71. **Question**: What is the formula for calculating the unit weight of soil?

   - Answer: The unit weight ($\gamma$) of soil can be calculated using the formula:
   $$\gamma = \frac{W}{V}$$
   where $W$ is the weight of the soil sample and $V$ is the volume of the soil sample.

72. Question: Define hydrology and its significance in civil engineering.

   - Answer: Hydrology is the study of the distribution, movement, and properties of water on Earth. It is significant in civil engineering for designing drainage systems, flood control measures, and water resource management.

73. **Question**: How do you calculate the bearing capacity of shallow foundations using the Meyerhof method?

   - Answer: The bearing capacity of shallow foundations using the Meyerhof method can be calculated using the formula:
   $$q_{ult} = c'N_c + \frac{qN_q}{F_s} + 0.5 \gamma BN_\gamma$$
   where:
   - $q_{ult}$ is the ultimate bearing capacity of the soil,
   - $c'$ is the effective cohesion of the soil,
   - $\gamma$ is the unit weight of the soil,
   - $B$ is the width of the foundation,
   - $N_c$, $N_q$, and $N_\gamma$ are bearing capacity factors, and
   - $F_s$ is the shape factor.

74. Question: Define pile foundation and its types.

   - Answer: A pile foundation is a type of deep foundation that transmits building loads to deeper, more stable soil or rock strata. Types of pile foundations include driven piles (such as steel H-piles and concrete piles) and drilled piles (such as bored piles and auger-cast piles).

75. **Question**: What is the formula for calculating the flow rate through an orifice?

   - Answer: The flow rate ($Q$) through an orifice can be calculated using the formula:
   $$Q = C_d \cdot A \cdot \sqrt{2gh}$$
   where:
   - $C_d$ is the discharge coefficient,
   - $A$ is the area of the orifice,
   - $g$ is the acceleration due to gravity, and
   - $h$ is the height of the fluid above the center of the orifice.

76. Question: Define compaction factor in concrete technology.

   - Answer: Compaction factor is a measure of the workability of concrete, indicating the degree to which freshly mixed concrete can be compacted effectively. It is determined by the ratio of the observed density of compacted concrete to the density of fully compacted concrete.

77. Question: What is the purpose of environmental impact assessment (EIA) in civil engineering projects?

   - Answer: Environmental impact assessment (EIA) evaluates the potential environmental effects of proposed projects or developments, ensuring compliance with environmental regulations and minimizing negative impacts on ecosystems, biodiversity, and human health.

78. Question: How do you calculate the shear force in a simply supported beam under a uniformly distributed load?

   - Answer: The shear force ($V$) in a simply supported beam under a uniformly distributed load can be calculated using the formula:
   $$V = \frac{wL}{2}$$
   where:
   - $w$ is the uniformly distributed load, and
   - $L$ is the span length of the beam.

79. Question: Define slump test in concrete technology.

   - Answer: The slump test is a standard method used to measure the consistency and workability of freshly mixed concrete. It involves filling a conical mold with concrete, compacting it, and then measuring the settlement (slump) of the concrete after removing the mold.

80. **Question**: What is the formula for calculating the flow velocity in an open channel using Manning's equation?

   - Answer: The flow velocity ($v$) in an open channel using Manning's equation is given by:
   $$v = \frac{1}{n} \cdot R^{2/3} \cdot S^{1/2}$$
   where:
   - $n$ is Manning's roughness coefficient,
   - $R$ is the hydraulic radius, and
   - $S$ is the slope of the channel.

81. Question: Define beam-column in structural engineering.

   - Answer: A beam-column is a structural member subjected to combined bending and axial loading. It is commonly used in building frames and other structural systems to transfer vertical loads and resist lateral loads.

82. Question: How do you calculate the lateral earth pressure on a retaining wall using Coulomb's theory?

   - Answer: The lateral earth pressure ($P$) on a retaining wall using Coulomb's theory can be calculated using the formula:
   $$P = K_p \cdot \gamma \cdot H$$
   where:
   - $K_p$ is the coefficient of active earth pressure,
   - $\gamma$ is the unit weight of soil, and
   - $H$ is the height of the retained soil above the base of the wall.

83. Question: What is the formula for calculating the principal stresses in a soil element under plane strain conditions?

   - Answer: The principal stresses ($\sigma_1$ and $\sigma_3$) in a soil element under plane strain conditions can be calculated using the formula:

   $$\sigma_1 = \frac{\sigma_x + \sigma_y}{2} + \sqrt{\left(\frac{\sigma_x - \sigma_y}{2}\right)^2 + \tau_{xy}^2}$$

   $$\sigma_3 = \frac{\sigma_x + \sigma_y}{2} - \sqrt{\left(\frac{\sigma_x - \sigma_y}{2}\right)^2 + \tau_{xy}^2}$$

84. Question: Define arch dam and its advantages.

   - Answer: An arch dam is a type of dam that curves upstream, resembling the shape of an arch. Its advantages include high strength, stability against thrust forces, and suitability for narrow canyon sites with strong foundation rock.

85. **Question**: How do you calculate the modulus of elasticity of concrete?

   - Answer: The modulus of elasticity ($E$) of concrete can be estimated using empirical formulas based on the compressive strength of concrete, such as:

   $$E = 4700 \sqrt{f_c}$$

   where $f_c$ is the compressive strength of concrete in psi.

86. Question: What is the purpose of a geosynthetic clay liner (GCL) in environmental engineering?

   - Answer: A geosynthetic clay liner (GCL) is used as a barrier in landfill liners, containment ponds, and other environmental applications to prevent the migration of contaminants into the surrounding soil and groundwater.

87. Question: Define time of concentration in hydrology.

   - Answer: Time of concentration is the time taken for runoff to travel from the farthest point of a watershed to the outlet. It is a critical parameter in hydrologic modeling for designing stormwater management systems and predicting flood peaks.

88. Question: What is the formula for calculating the flow rate through a rectangular weir?

   - Answer: The flow rate ($Q$) through a rectangular weir can be calculated using the Francis

91. Question: Define slump flow test in concrete technology.

   - Answer: The slump flow test is a modified version of the standard slump test used to assess the workability and flowability of self-compacting concrete (SCC). It measures the spread of concrete after being placed in a slump cone without any external forces.

92. Question: What is the purpose of a geotechnical baseline report (GBR) in tunneling projects?

   - Answer: A geotechnical baseline report (GBR) provides a comprehensive overview of anticipated subsurface conditions and geological challenges along the tunnel alignment. It serves as a reference document for assessing variations from expected conditions during tunnel construction.

93. Question: How do you calculate the settlement of a shallow foundation on granular soil using the elastic settlement method?

   - Answer: The settlement ($S$) of a shallow foundation on granular soil using the elastic settlement method can be calculated using the formula:
   $$S = \frac{q}{E} \cdot \left(\frac{1}{1+\nu}\right) \cdot H$$
   where:
   - $q$ is the applied pressure,
   - $E$ is the modulus of elasticity of the soil,
   - $\nu$ is Poisson's ratio of the soil, and
   - $H$ is the height of the foundation.

94. Question: Define effluent in wastewater treatment.

   - Answer: Effluent is the treated or untreated wastewater discharged from a wastewater treatment plant or industrial process into the environment, typically into rivers, lakes, or oceans. Effluent quality is regulated to minimize environmental impact and protect public health.

95. Question: What is the formula for calculating the modulus of subgrade reaction for a rigid pavement?

   - Answer: The modulus of subgrade reaction ($k$) for a rigid pavement can be calculated using the formula:
   $$k = \frac{P}{\delta}$$
   where:
   - $P$ is the load applied to the pavement, and
   - $\delta$ is the deflection of the pavement surface.

96. Question: Define hydraulic conductivity in hydrogeology.

   - Answer: Hydraulic conductivity (also known as permeability) is a measure of a material's ability to transmit fluids through its pore spaces under a hydraulic gradient. It is a fundamental property in groundwater flow and aquifer characterization.

97. Question: How do you calculate the moment of resistance of a reinforced concrete section?

   - Answer: The moment of resistance ($M_R$) of a reinforced concrete section can be calculated using the formula:
$$M_R = A_s \cdot f_y \cdot d \cdot (1 - \frac{d}{2c})$$
where:
   - $A_s$ is the area of tension reinforcement,
   - $f_y$ is the yield strength of the reinforcement,
   - $d$ is the effective depth of the section, and
   - $c$ is the distance from the extreme compression fiber to the centroid of tension reinforcement.

98. Question: Define hydraulic jump in open channel flow.

   - Answer: A hydraulic jump is a phenomenon that occurs when supercritical flow rapidly transitions to subcritical flow, resulting in a sudden increase in water depth and energy dissipation. It is commonly observed downstream of weirs, spillways, and obstructions in open channels.

99. Question: What is the purpose of a geotechnical centrifuge in soil mechanics?

   - Answer: A geotechnical centrifuge is a laboratory instrument used to simulate the effects of gravity on soil behavior at small scales. It allows researchers to study complex soil-structure interactions, slope stability, and foundation performance under simulated gravitational conditions.

100. Question: How do you calculate the uplift pressure on the base of a submerged structure?

   - Answer: The uplift pressure ($P_u$) on the base of a submerged structure can be calculated using the formula for hydrostatic pressure:

   $$P_u = \gamma_w \cdot H$$

   where:

   - $\gamma_w$ is the unit weight of water, and
   - $H$ is the depth of water above the base of the structure.

101. Question: Define specific gravity in soil mechanics.

   - Answer: Specific gravity is the ratio of the density of a substance to the density of a reference substance (usually water). In soil mechanics, specific gravity is used to characterize the properties of soil particles and determine soil classification.

102. Question: What is the formula for calculating the flow rate through a venturi meter?

   - Answer: The flow rate ($Q$) through a venturi meter can be calculated using the formula:

   $$Q = C_d \cdot A_1 \cdot \sqrt{2gH}$$

   where:

   - $C_d$ is the discharge coefficient,
   - $A_1$ is the area of the throat of the venturi, and
   - $H$ is the pressure head difference between the throat and upstream section.

103. Question: Define critical path method (CPM) in project management.

   - Answer: The critical path method (CPM) is a project management technique used to schedule and manage complex projects by identifying the longest sequence of dependent activities (the critical path). It helps project managers allocate resources, track progress, and identify potential delays.

104. Question: How do you calculate the volume of earthwork for a triangular prism?

   - Answer: The volume ($V$) of earthwork for a triangular prism can be calculated using the formula:

   $$V = \frac{1}{2} \cdot b \cdot h \cdot L$$

   where:
   - $b$ is the base width of the triangular cross-section,
   - $h$ is the height of the triangular cross-section, and
   - $L$ is the length of the prism.

105. Question: Define ultimate bearing capacity in foundation engineering.

   - Answer: Ultimate bearing capacity is the maximum pressure or load that a foundation soil can support without undergoing shear failure. It is a critical parameter for designing safe and stable foundations for structures.

106. Question: What is the formula for calculating the shear stress in a fluid flow through a pipe?

AnswAnswer: The shear stress ($\tau$) in a fluid flow through a pipe can be calculated using the formula for viscous flow:

$$\tau = \frac{\mu}{r} \cdot \frac{du}{dy}$$

where:
- $\mu$ is the dynamic viscosity of the fluid,
- $r$ is the radial distance from the pipe axis, and
- $\frac{du}{dy}$ is the velocity gradient in the radial direction.

107. Question: Define retention time in wastewater treatment.

- Answer: Retention time is the duration that wastewater remains in a treatment process or reactor, typically measured in hours or days. It is a critical parameter in wastewater treatment design, influencing the efficiency of physical, chemical, and biological processes for pollutant removal.

108. Question: How do you calculate the time of concentration in hydrology?

- Answer: The time of concentration ($T_c$) in hydrology can be calculated using empirical equations such as the Rational Method or the Kirpich Equation, which consider factors such as watershed size, slope, land use, and rainfall intensity.